Eight frowning clowns
are in a stew.
Their balancing act
is hard to do.

Seven are low;
one's way up high —
that's not right
but they don't know why.

They start again
with six and two.
No, no, no,
that just won't do!

Next the clowns
try five and three —
that's still not how
it's meant to be!

Then suddenly
they give a shout.